Shoshin

Shoshin

The System for AI-Powered Business Transformation

Scott Duffy

Shoshin

The System for AI-Powered Business Transformation

Scott Duffy

ISBN: 978-1-968149-11-6

Joint Venture Publishing

The Millionaire Mentor, Inc

JOINT VENTURE

Printed in the United States of America

"Beginner's luck is not an accident.
It's a system."

— *Scott Duffy*

PREFACE

Transformation as a System

We're living through the **fastest period of transformation** in human history. The difference between those who thrive and those who fall behind isn't access to technology. It is the ability to transform continuously.

This book is a **blueprint for turning transformation into a system** so that leaders and their teams have the mindset, structure, and rhythm to stay ahead of every wave that follows.

TRANSFORMATION SPIRAL

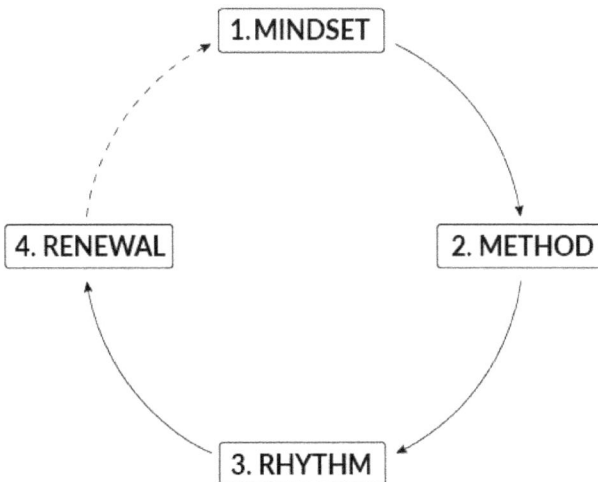

- 1. MINDSET
- 2. METHOD
- 3. RHYTHM
- 4. RENEWAL

Dedication

To Lily and Lexi

I am so proud to be your dad.

AUTHOR'S NOTE

The Right Book at the Right Time

On the Books and Ideas That Shaped This One

I've always been a voracious learner. Every major transformation in my life began with a book.

In the '80s, when I first entered the world of training, it was *Personal Power* by Tony Robbins, *The Psychology of Winning* by Denis Waitley, *The 7 Habits of Highly Effective People* by Stephen Covey, and *How to Win Friends and Influence People* by Dale Carnegie. Those books taught me that mindset is the foundation of success.

In the '90s, during the rise of the Internet, it was *Crossing the Chasm* by Geoffrey Moore, *The Innovator's Dilemma* by Clayton Christensen, *Competing for the Future* by Gary Hamel and C.K. Prahalad, and *Permission Marketing* by Seth Godin. They showed me how technology changes everything - and how timing, not talent, often determines success.

In the 2000s, as an entrepreneur building and scaling businesses, it was *The Lean Startup* by Eric Ries, *Zero to One* by Peter Thiel, *The Tipping Point* by Malcolm Gladwell, and Exponential Organizations by Salim Ismail. Those books helped me understand that iteration and experimentation are the new competitive advantages.

In recent years, during the AI revolution, it's been *The Singularity Is Near* by Ray Kurzweil, *The Future Is Faster Than You Think* by Peter Diamandis and Steven Kotler, *Superintelligence* by Nick Bostrom, and *Digital Dharma* by Deepak Chopra. These books revealed how exponential technologies are reshaping the world faster than we can process.

And then there's the group of books that grounded my thinking about transformation itself: *Zen Mind, Beginner's Mind* by Shunryu Suzuki, *Eleven Rings* by Phil Jackson, *The Creative Act: A Way of Being* by Rick Rubin, and *Breaking the Habit of Being Yourself* by Joe Dispenza. These works reminded me that while technology accelerates the pace of change, it's mindset, creativity, and self-awareness that determine how we handle it.

For me, each of these books shaped the way I see transformation. Together, they helped me create the methodology you'll find in this book: **a system for turning curiosity into speed, and speed into measurable transformation.**

Shoshin is the sum of everything I've learned across those decades. It's the mindset, models, and methods that make transformation repeatable.

My hope is that this book becomes your "right book at the right time," the one that helps you stay curious, stay adaptable, and keep beginning, no matter how fast the world changes.

— Scott Duffy

HOW TO USE THIS BOOK

Read, Reflect, Apply

This book is divided into two parts.

Part I: The Shoshin System.

In this section, you will learn how transformation works, why organizations get stuck, and how to build the mindset and structure to move faster. Each chapter in Part 1 ends with a Practice: a short exercise designed to help you apply the ideas immediately.

Part II: The Shoshin Playbook.

This is your toolkit. It contains the full set of exercises, templates, and workflows used by leadership teams to build AI-ready organizations. It also includes a QR code linking to digital worksheets your team can use to collaborate and track progress.

Read with your leadership team. Discuss weekly. Apply one practice at a time.

Start small. Stay curious. Keep beginning.

TABLE OF CONTENTS

PART I: The Shoshin System

PART II: The Shoshin Playbook

PART I

The *Shoshin* System

CHAPTER 1

START: The Mindset of a Beginner

And How It Helps Winners Win

I grew up near the beach in Southern California, and I've always believed the ocean teaches transformation better than any boardroom ever could.

When a new set of waves starts to rise, most people in the water freeze. They turn toward shore, sprint for safety, and end up exactly where the wave breaks—right in the chaos.

The experienced surfer does the opposite. They paddle straight at the wall of water and, at the last second, push their board down and slide underneath. Surfers call that a duck dive. For a few seconds, the world is turbulence above and stillness below. Then you come out the other side, in calm water, ready for the next wave.

Business disruption works the same way. When markets shift or technology changes, most companies retreat. They wait for the water to calm. But delay multiplies turbulence. There is no path around transformation, only through it.

Leaders who dive early surface first.

Why Winners Win

I have spent most of my career studying people at the top of their field, asking one question:

Why do some people keep winning, no matter how much things change around them?

Years ago, I spent time with the Los Angeles Lakers during one of their championship runs under Phil Jackson. By that time, Jackson had already won more championships than any coach in professional sports history. He was also the only coach to lead teams built around both Michael Jordan and Kobe Bryant, two of the greatest players to ever play the game.

I kept asking myself the same question: How did Jackson help generational players like these keep winning, even as the game evolved, the league shifted, and the competition intensified?

At first glance, the answer seems obvious. Michael and Kobe had it all. Talent. Drive. Relentless work ethic. They demanded excellence from themselves and everyone around them. They were wired for greatness. It's easy to assume Jackson simply got more out of extraordinary talent.

But talent alone doesn't explain sustained winning.

History is full of brilliant players and organizations that reached the top once, then lost momentum when conditions changed.

What fascinated me wasn't that Michael and Kobe became great. It was that they stayed great—season after season, era after era—while the world around them kept evolving.

This is what separated Jackson as a leader.

He didn't rely on motivation or talent. Instead, he built a system for sustaining greatness.

Through his leadership philosophy, rituals, and relentless emphasis on learning, Jackson reinforced a discipline elite performers often lose as they accumulate wins: the willingness to approach each game as if it were their first.

In Zen Buddhism, that discipline is called *Shoshin*. It translates to *"beginner's mind."*

Jackson didn't create greatness. He protected it.

Every time his teams stepped onto the court, they were asked to let go of the last win, the last loss, and the last identity. Each game became a new problem to solve, not a performance to defend.

That behavior reveals the pattern behind sustained excellence.

Experts protect what they know. Beginners explore what they don't. Winners stay beginners.

BEGINNER VS EXPERT CONTRAST

EXPERT MINDSET *(CLINGS TO WHAT WORKS)*	BEGINNER MINDSET *(SEES WITH FRESH EYES)*
Defends what works	Questions assumptions
Waits for certainty	Acts with curiosity
Optimizes the past	Explores what's next
Avoids discomfort	Learns through uncertainty
Protects expertise	Builds adaptability

For Michael and Kobe, this wasn't a philosophy.

It was their competitive edge.

In every era of transformation, new leaders emerge and old ones fade. The winners who endure share one defining trait: a beginner's mindset. They approach change with curiosity instead of rigidity. They question assumptions. They let go early. And they begin again with small deliberate actions that build momentum..

When What Made You Great Starts Working Against You

A CEO I worked with had built a category-leading company over more than a decade. At its peak, everything worked. His instincts were sharp. His routines were disciplined. His decisions were fast and confident. The habits that defined his leadership had been rewarded again and again by the market.

Then the environment changed.

New competitors emerged. Technology shifted. Customer expectations evolved. But instead of adapting, he doubled down on what had always worked. He optimized harder. He enforced the same playbook with more rigor. He trusted the instincts that had made him successful.

The results got worse.

Revenue slowed. Teams lost momentum. Market share slipped. What once felt like mastery now felt like friction, but no one could figure out why.

When he finally asked for help, we didn't start by fixing the business. **We studied patterns.**

One team documented everything he did when the company was winning: how decisions were made, how priorities were set, where time was spent, what behaviors were rewarded. Another team studied leaders in the same industry who were winning now. Not the loudest or most visible, but the fastest-adapting.

We put both approaches on a single page.

The contrast was undeniable.

What had once been his competitive advantage had quietly become his constraint. The very behaviors that **created success** in one era were **preventing adaptation** in the next.

The turning point wasn't a new strategy. It was a moment of openness.

He looked at the page for a long time and finally said, "If I were starting this company today, I wouldn't do half of this."

That was Shoshin in action.

Within months, the company changed course. Mindset changed. Experimentation began. Learning replaced certainty. Performance stabilized, then accelerated. The leader didn't abandon his experience, but he stopped letting it dictate the future.

What had once been his formula for sustained success had turned into dead weight, pulling his business under water. Momentum returned only when he loosened his grip and embraced the challenge with a beginner's mind.

Before any system can work, leaders must remove the mental constraints that slow adoption, delay action, and protect outdated certainty.

Leaders don't fail because they lack intelligence or experience. They fail because they cling to certainty during times of change. What once created success quietly becomes identity. Identity becomes comfort. And comfort becomes the enemy of adaptation. Shoshin is not about thinking less like an expert. It's about recognizing when expertise has turned into a constraint.

Transformation doesn't start with the market, the team, or the technology. It starts with the leader's willingness to question their own certainty.

Momentum Math

Once you understand that transformation begins with you, the next question becomes simple:

How do you create momentum that lasts?

Motivation will not get you there. Motivation is emotion, and it fades. Momentum is physics, and it compounds.

Two friends once made a bet about who could get in better shape. The first friend was perfect for six days. Then Sunday arrived: pizza, cake, a break in the pattern, and he was back to zero. Every week he repeated the cycle. Every Monday became another restart.

Meanwhile, the second friend kept doing the right things for 30 straight days. He lost 10 pounds. His inconsistent friend lost none. The friend who won the bet had the power of momentum.

The same math applies in business. Ten minutes a day doing the right things beats 10 hours once a week. One small action performed consistently builds momentum and creates exponential return.

Focus × Consistency = Momentum

Small, simple actions such as testing one AI workflow, updating a single process, or teaching one teammate outperform massive quarterly initiatives that never sustain. Even 10 minutes saved per day using a basic AI tool saves more than 43 hours per year—the equivalent of a full workweek.

Momentum isn't luck. It's the physics of persistence.

HOW TO BUILD MOMENTUM

Principle	Enterprise Application
Start Small	Run small, regular micro-pilots.
Measure Quick	Evaluate results rapidly; keep or cut.
Share Learning	Turn wins into new standards.

The "Day One" Reset

Two founders had built a successful company over 15 years, but the market was shifting fast. New competitors were stealing share. One partner wanted to evolve; the other insisted on doubling down on what had always worked. Revenue slid. Tension rose. They were close to splitting.

The restless partner called me and said, "I'm done. I'll start a new company, a direct competitor, right across the street."

I said, "Before you do that, let's try something. Schedule a meeting with the three of us."

When we met, I suggested we conduct a thought experiment. **"Pretend it's day one. Today's market, today's problem, no baggage. What would you build?"**

At first, they looked at each other like I was crazy. Then the ideas started flowing. By the end of the meeting, they had a completely new model: fresh pricing, digital channels, a new client experience. The room shifted from frustration to energy. They stopped defending the past and started designing the future.

That's Shoshin at work; curiosity over certainty. Within five years, the company tripled in size. The strategy wasn't new. The mindset was.

Renewal starts with subtraction. Ask:

"If we started building this today, what would we do differently?"

Shoshin Practice #1:
Start Small, Start Now (The Daily Reset)

Every morning is another wave. A quick reset keeps you in beginner's mind. Ten minutes a day, every day, is enough to build momentum.

Ask yourself three questions:

1. What beliefs about our business have served us well, and do they still serve us today?

2. If I were starting this company or department from scratch today, what would I build differently?

3. What's one small action I can take right now to start building momentum?

Begin each morning with these questions for a week. They will keep you curious, help you surface assumptions, and move you from intention to action.

Starting is the hardest part, but movement creates clarity. In the next chapter, we will look at why every wave of innovation follows a familiar pattern and how understanding that pattern helps you recognize the signals of transformation before others see them

a familiar pattern and how understanding that pattern helps you recognize the signals of transformation before others see them.

CHAPTER 2

HISTORY:
How to Recognize the Signals of Transformation

And What AI Teaches Us About Every Wave That Came Before

Every Revolution Looks the Same

Every few generations, the world resets.

Fire. The wheel. The printing press. Electricity. The Internet. AI. Each transformation felt unique to the people living through it. But viewed from a distance, they all follow the same rhythm:

Curiosity → Fear → Experimentation → Adoption → Reinvention

AI is simply the newest wave in a thousand-year pattern. The difference now is tempo. Where the Industrial Revolution unfolded over decades, AI is compressing change into months.

Technology moves fast. People move slow. Leaders bridge the gap.

This chapter is about recognizing the early signals that another wave has begun and building a culture that can move with it instead of being crushed beneath it.

The Four Modern Waves of Innovation

Over the last four decades, I've lived through four modern innovation waves. Each felt disruptive at the time. In hindsight, they were chapters in a single story: humanity's ability to learn and adapt faster.

FOUR MODERN WAVES OF INNOVATION

Wave	Era	Defining Shift	Examples	Lesson for Leaders
Information Wave	Mid-1980s to mid '90s	Knowledge becomes the product.	Tony Robbins, VHS training, desktop publishing, the rise of "the knowledge economy"	Value moved from what you build to what you know.
Internet Wave	Mid-'90s to 2000s	Connectivity becomes universal infrastructure.	Netscape, Yahoo!, MSN, eBay, Amazon	Speed + access redefined markets overnight. If you weren't connected, you weren't visible.
Platform Wave	2010s	Scale comes through ecosystems and data.	Facebook, LinkedIn, Salesforce, mobile apps, SaaS	Ecosystems beat products; networks beat features.
AI Wave	2020s to now	Intelligence becomes leverage.	ChatGPT, generative AI, automation, robotics	Humans + machines outperform either alone. Transformation now means learning to think with AI.

Each wave felt shorter than the one before. As adoption accelerates, so does opportunity. The pace of change that was once linear has become exponential. **The leaders who learn fastest create the biggest advantage.**

When Johannes Gutenberg invented the printing press around 1440, it took more than 50 years before books became common across Europe. In the age of AI, 50 weeks can reshape an industry.

Transformation used to be a once-in-a-career event. Now it's a quarterly discipline.

You can't slow the waves, but you can learn to ride them.

How to Recognize the Signals of Transformation

Every revolution starts as background noise. By the time most people realize what's happening, the new world is already here.

Over time, I've learned to look for three signals that always appear early:

1. The Toy Turns Useful.

The first version of every great technology looks like a toy. Early websites were digital brochures. The first smartphones were novelties. Early AI wrote clumsy sentences. Then one day, the "toy" solves a real problem better, faster, and cheaper. Everything changes.

2. The Skeptics Run Out of Arguments.

At first, experts explain why it won't work. Then, they quietly start using it themselves. Once the objections dry up, adoption accelerates.

3. The Next Generation Stops Asking Permission.

Nothing signals a shift faster than young talent moving without approval. Once the next generation adopts a technology as their default, the rest of the world is already behind.

These signals appear in every industry, every time. Leaders who notice them early act while others wait for clarity.

That's the essence of Shoshin: curiosity before certainty.

A Brief History of AI

Artificial Intelligence isn't new. The term was coined at the Dartmouth Summer Research Project in 1956, when researchers set out to build machines that could learn. Progress came in bursts: optimism in the '60s, slow winters in the '70s and '80s, breakthroughs in the 2010s with deep learning.

Then, in November 2022, everything changed. ChatGPT made AI accessible to everyone. Overnight, the conversation shifted from if AI would matter to how fast it would transform everything.

Within months, investment soared. Organizations raced to experiment while others froze, waiting for clarity that never comes. The psychology was ancient:

Curiosity → Fear → Experimentation → Adoption → Reinvention

The same pattern that followed the printing press, electricity, and the Internet was repeating again, only faster.

The leaders moving now aren't the ones with all the answers. They're the ones willing to learn in public.

The Transformation Bridge

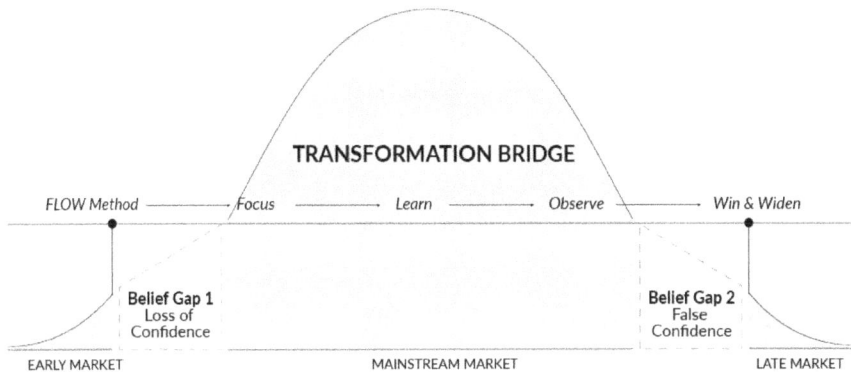

TRANSFORMATION BRIDGE

FLOW Method ———→ Focus ———→ Learn ———→ Observe ———→ Win & Widen

Belief Gap 1
Loss of
Confidence

Belief Gap 2
False
Confidence

EARLY MARKET   MAINSTREAM MARKET   LATE MARKET

The Transformation Bridge is a system for continuous transformation. Powered by the *FLOW* Method, it moves organizations from one cycle to the next—without stalling in uncertainty or locking into false confidence.

In 1991, Geoffrey Moore published Crossing the Chasm, one of the most influential business books ever written. Moore's core insight was that new technologies are not adopted in a straight line. Markets move through predictable phases of enthusiasm, skepticism, and eventual maturity.

His book codified the now-standard vocabulary for market adoption: **Innovators → Early Adopters → Early Majority → Late Majority → Laggards.** It also identified a dangerous midpoint he called the chasm.

What is the chasm?

It's the point where **early excitement meets operational reality.**

Where **expectations exceed results.**

Where **promising projects get dropped** because organizations don't yet know how to integrate a new capability into real workflows.

Moore's work is essentially a **textbook for product creators** who want to learn how to design, launch, and scale new technologies so they successfully enter the mainstream.

The Transformation Bridge is different.

It's designed for the organizations on the receiving end; the ones that must adopt these emerging technologies and manage the emotional, operational, and cultural changes required to stay ahead.

Moore explains the **pace of adoption.** The Transformation Bridge explains the energy of belief. Because transformation doesn't fail when technology is immature. It fails when belief collapses.

The Three Stages of Transformation

Inside organizations, transformation typically moves through three phases:

1. **Early Market:** Curiosity, experimentation, early wins.
2. **Mainstream Market:** Proof, repeatability, scalable value.
3. **Late Market:** Skepticism, resistance, eventual acceptance.

At first glance, this looks similar to a traditional adoption curve. So what's different about the Transformation Bridge versus other models? Three things:

1. Audience

Most models, including Moore's, are built for product developers bringing new technologies to market. The Transformation Bridge is built for the organizations adapting to those technologies and navigating **the human side of change.**

2. Belief Gaps

The Transformation Bridge introduces the idea that transformation breaks down at two recurring points, early and late, where belief collapses and momentum stalls.

3. System + Straight Line

Transformation isn't a curve. **It's a straight-line operating system** that becomes part of culture. Rather than riding the emotional waves of the market, it keeps teams moving seamlessly from new technology to new technology and from innovation to innovation.

These differences shift the entire conversation from technology readiness to human psychology and **organizational behavior.**

The First Belief Gap: Early Market → Mainstream

The Early Market begins energized. Teams experiment. Leaders explore. Budgets stay flexible. Curiosity drives momentum. But a predictable pattern emerges:

- Companies overinvest too early.
- Experiments don't work as expected.
- Integration is harder than planned.
- ROI takes longer than projected.
- Leaders lose patience.
- Teams lose conviction.

This is where most transformation efforts stall.

Moore called this area **the chasm.**

Gartner later described a similar moment as the Trough of Disillusionment, where organizations abandon promising technologies after early setbacks.

But in transformation, something deeper is happening.

It's not a technology failure.

It's a belief failure.

Projects don't die because the solution is incapable. They die because teams don't believe, or don't see, a way through the messy middle.

This is the **First Belief Gap.**

What happens when belief collapses? Early pilots pause, budgets freeze, experimentation stops, momentum disappears, and everyone waits for "proof" that only appears after the gap.

The irony is profound. **The results everyone wants are waiting just beyond the gap.**

The Second Belief Gap: Mainstream → Late Market

Most models stop after the first gap. Transformation doesn't.

The Second Belief Gap appears on the far side of the curve, between the Mainstream and the Late Market. This group isn't wrestling with technical failures. They're wrestling with identity, comfort, and certainty.

- Their beliefs sound like this:
- "This doesn't apply to our team."
- "We've always done it this way."
- "Let's wait until it's fully proven."
- "Our customers won't care."

- "Why change? Everything is working fine."

This isn't disillusionment. **It's entrenched certainty.**

What finally pulls the Late Market forward? **Undeniable proof + the threat of loss.**

They move when:

- competitors outpace them.
- customers demand new capabilities.
- partners require new standards.
- the cost of standing still becomes greater than the cost of change.
- *yes, their job security is at risk.*

The Late Market doesn't move because the technology improves. It moves because the world around them provides **irrefutable data** and forces belief to catch up.

This is the **Second Belief Gap.**

Why the Transformation Bridge Is Not a Curve

Moore's model is a bell curve because markets move through external dynamics: hype, disappointment, recovery, maturity.

The Transformation Bridge is something entirely different: **A straight line running directly through the center of the curve.**

It represents stability, discipline, and a controlled emotional state.

It doesn't rise with hype. It doesn't dip with disappointment. It doesn't spiral into late-stage skepticism.

It holds steady.

Why is the Transformation Bridge a straight line? Because transformation requires shielding the organization from external volatility and replacing it with an **internal operating rhythm** that produces reliable, compounding progress.

The Transformation Bridge keeps organizations from being pulled off course by the emotional momentum of the curve around them.

- Too high too early: Overconfidence and reckless bets.
- Too low at mid-curve: Frustration, freezes, and cancellations.
- Too rigid too late: Identity-driven resistance.

The Transformation Bridge protects against all three. It keeps teams from getting too high, too low, too reactive, and too hesitant.

And it creates the conditions for **continuous learning, growth, adaptation, and transformation.**

In an AI era of constant, never-ending change, the ability to stay centered - to test, observe, implement what works, and keep moving is what separates the winners from those left behind.

FLOW Is the Engine That Keeps You on the Transformation Bridge

Why does the Transformation Bridge stay straight while the world around it moves like a bell curve? Because *FLOW* eliminates the need for dramatic swings in belief.

FLOW gives transformation a **sustainable rhythm:**

- Short cycles prevent early overinvestment.
- Small bets reduce risk and build confidence.
- Visible wins create proof where belief is weakest.
- Structured reflection removes ego and accelerates correction.
- Consistent cadence prevents emotional spikes or crashes.

FLOW **keeps belief stable.** It provides

- just enough evidence early
- just enough refinement mid-curve
- just enough structure late

to cross both belief gaps smoothly and predictably.

The Essence of the Transformation Bridge

Most organizations fall into the gaps because they mirror the emotional momentum of the bell curve. Leaders using the

Transformation Bridge behave differently:

- They avoid early hype.
- They avoid mid-curve disappointment.
- They don't wait until it's safe.
- They don't freeze when discomfort rises.
- They don't cling to the old way when the world moves on.

They maintain belief—not as emotion, but as a system. And that system carries them from one transformation cycle to the next without collapse, stall-out, or reboot.

The curve explains what **the world does.**

The Transformation Bridge explains what great organizations do.

Why the Transformation Bridge Changes Everything

Most organizations don't fail at transformation because they lack intelligence, talent, or resources. They fail because they misinterpret what's happening when momentum slows.

They see early friction and assume **the technology is wrong.**

They feel mid-cycle discomfort and **assume the timing is off.**

They encounter late resistance and **assume their people aren't ready.**

In reality, they're standing inside predictable belief gaps without a system to carry them across. The Transformation Bridge changes the outcome by **reframing the problem leaders think they're solving.**

It teaches leaders to stop reacting emotionally to where they are on the curve and start operating deliberately through it. It replaces guesswork with pattern recognition. Panic with cadence. Hope with proof.

Organizations that use the Transformation Bridge don't move faster because they take bigger risks. They move faster because they **don't lose belief at the wrong moments.**

They **stay centered** when others get excited.

They **stay disciplined** when others get discouraged.

They **keep moving** when others pause to wait for certainty.

That stability compounds.

Once leaders understand the Transformation Bridge, transformation stops feeling chaotic. It becomes navigable. Teams know what phase they're in, what doubts to expect, and what kind of proof is required next.

That clarity is the real advantage.

The Transformation Bridge does not guarantee success, but it dramatically reduces unnecessary failure. It's the foundation that allows opportunity to be seen clearly, experiments to be run safely, and momentum to be sustained long enough for results to appear.

And that is **why beginners win.**

Because when others wait for confidence, beginners build it one small proof point at a time.

Shoshin Practice #2: See the Pattern

Ask yourself three questions:

1. What are we holding onto because it's always worked?
2. Who inside our organization already sees the next wave and is experimenting?
3. What proof points can we create this quarter to help others cross the Transformation Bridge to Transformation?

Transformation gets easier once you can see the pattern and trust the signals.

History doesn't repeat itself. It accelerates. Once you learn to read the signals, you stop fighting the wave and start riding it.

In the next chapter, we'll explore where opportunity hides inside these moments of uncertainty, and why those who move with curiosity instead of certainty are the first to find it.

CHAPTER 3

OPPORTUNITY:
How the Beginner Finds Advantage
And Why AI Fluency, Not Training, is the Holy Grail

Every Transformation Opens a Door

Every transformation opens a door that most experts cannot see. It rarely looks clean or convenient; in fact, it often looks unclear, unproven, and unpredictable. But that's exactly where opportunity hides.

Leaders who practice Shoshin, the beginner's mind, see what others overlook. They don't wait for certainty. They act with curiosity. Every disruption becomes an open field for discovery.

One global company tried to build an enterprise-wide AI system from scratch. The project was big, ambitious, and siloed. Budgets poured in, but communication broke down and momentum stalled.

Meanwhile, a small team in marketing started experimenting. They used ChatGPT to summarize reports and create briefs. They shared what worked and, within weeks, other departments followed. Their curiosity created a connection. Their results created momentum. The transformation went viral across the company.

That's the difference. The first team tried to build a solution in isolation. The second team practiced Shoshin and applied the FLOW principles:

Focus • Learn • Observe • Win and Widen

If you're wondering where to start when leading through transformation, start with enterprise-wide education. This is the lesson:

CURIOSITY EQUATION

$$\frac{CURIOSITY \times SPEED\ TO\ LEARNING \times IMPLEMENTATION}{= TRANSFORMATION}$$

Curiosity x Speed to Learn x Implementation = Transformation

The organizations that win don't bet big. They learn fast. They protect the downside, act quickly, and share what they learn. That combination of velocity and mitigated risk turns uncertainty into advantage.

Education is Where Transformation Begins

Always begin with education. When everyone understands the tool, fear drops, imagination rises, and ideas surface from every level of the company.

But training in the age of AI requires a different approach. It's no longer enough to simply learn a tool. Traditional software training focused on functionality and what buttons to push. You learned it once and were done.

With AI, the tools evolve too quickly for that to matter. What matters now is fluency: the ability to think, communicate, and problem-solve with AI. Specific tools will change; the system for learning, thinking, and working with them will not.

The New Rules of Training in the Age of AI

Training is no longer just about sharing information. It's about teaching new ways to think.

If you have 500 AI-trained employees, they will always lose to 10 AI-fluent ones.

Fluency means understanding how to collaborate with intelligence itself. It is knowing how to frame problems, challenge assumptions, and communicate with AI to achieve clarity and speed.

Here's a simple four-step model for developing AI fluency:

1. **Pick One Tool.**

In the early stages of a new market, focus on one foundational model or suite (ChatGPT, Claude, Gemini, Copilot, Canva for design, Midjourney for creative, or Harvey for legal) and identify where it can add real value.

2. **Think Like a Problem Solver.**

Learn to define the problem clearly, break it into parts that make it easy for AI to digest and respond quickly, and focus on measurable outcomes.

3. **Think Like a Critical Thinker.**

Challenge assumptions, test reasoning, and validate outputs. Train AI to examine your logic, not agree with you.

4. **Learn to Communicate with AI.**

Develop precise prompting skills. Define role, instructions, and context. Describe the desired output with clarity.

AI fluency requires both tool competence and the Shoshin mindset. Build these capabilities into every workshop, simulation, and project review. Together, they create an AI-fluent workforce that adapts as quickly as the technology itself.

From AI Fluency To Real Advantage

After you build AI fluency—knowing what AI can do, how to prompt it, and where it breaks—the next step is to master problem-solving and critical thinking.

AI is an amplifier, not a compass. It can generate options at lightning speed, but it can't reliably choose the right problem to solve or tell you whether an answer is actually true, complete, and relevant in your context.

Problem-solving ensures you're aiming at the right target with measurable outcomes.

Critical thinking ensures the output holds up under pressures such as assumptions, bias, tradeoffs, and evidence.

The Power of Problem Solving

Problem-solving is the ability to **define the right problem before trying to solve it.** In an AI-powered environment, this matters more than ever. AI will confidently solve whatever problem you give it, whether it's the right one or not.

Strong problem-solvers translate vague challenges into clear, structured questions with measurable outcomes.

For example, instead of saying, "Customer service is too slow," a team asks:

"How can we reduce average response time by 25% using automation without lowering customer satisfaction?"

To build problem-solving capability

- Teach simple frameworks like **root-cause analysis,** which means repeatedly asking "Why?" until you identify the real constraint instead of treating symptoms.

- Train teams to reframe issues from **symptoms to systems** by asking, "What's actually causing this—and where does it originate in the workflow?"

- Measure progress by the **quality of the problem definition,** not just the speed of the solution. If the problem is unclear, the output will be too.

Kickstarting Critical Thinking

Critical thinking is the ability to evaluate outputs, challenge assumptions, and test logic in real time. In the age of AI, this means resisting the temptation to accept polished answers at face value.

AI is designed to be helpful—and that often means it agrees with you. Critical thinkers push past agreement to ask whether the response is complete, unbiased, and contextually sound.

To strengthen critical thinking:

- Train teams to ask, **"What's missing from this answer?"** After AI generates a plan or recommendation, identify assumed data, missing context, or untested constraints.
- Use **structured debate** to expose weak reasoning. Have one group defend the AI's conclusion while another challenges it to surface gaps, risks, or alternative interpretations.
- Evaluate outputs based on **reasoning quality,** not surface-level correctness. Did the team question bias, evidence, and relevance—or simply accept a confident-sounding response?

The best leaders treat alignment as a **learning mechanism,** not a formality. After major decisions or AI-assisted projects, ask the one question I ask my editors:

"Do you agree with this or not—and why?"

That question forces people to articulate reasoning, surfaces disagreement early, and prevents teams from mistaking silence or AI confidence for true alignment.

The Beginner's Playbook for Transformation: Small, Fast Tests

Transformation isn't one big project; it's a series of small, fast tests run safely inside guardrails. Velocity with mitigated risk beats bold one-offs every time.

When we teach the Beginner's Playbook, we start with five rules:

1. Clearly define each test.

Identify scope, budget, goal, and success metric(s); for example, time savings, cost savings, improved quality, more efficient workflows, higher customer satisfaction, or better data accuracy.

2. Mitigate risk at every step.

Track progress, contain cost, set decision gates.

3. Run experiments in parallel.

Multiple small tests compound learning.

4. Speed beats certainty.

Perfection is the enemy of progress.

5. Document and share.

Turn learning into organizational memory.

Aim for a 10% improvement every 90 days in things like speed, quality, cost, customer satisfaction, or efficiency. Small gains compound into meaningful transformation.

Use proven foundational tools before you try to invent your own. An MIT study found that 95% of AI projects fail. The most common thread is companies trying to build custom systems instead of leveraging tested, stable, scalable platforms. Innovation should compound, not collapse due to complexity.

Communicating Transformation

AI changes technology, but it also changes emotions. Leaders need a communication strategy that turns fear into curiosity. Each stakeholder sees AI through a different lens.

Boards: Focus on competitiveness and risk.

"We're not automating people; we're ensuring we never fall behind those who do."

Investors: Link AI to profit and scale.

"We're improving productivity by 20% per cycle while de-risking long-term disruption."

Teams: Address fear directly.

"Some jobs will change; our mission is to help you lead that change."

"AI may change every role, but it won't replace people who learn to use it."

Then, back the words with action: training, transparency, and regular updates. When employees see small wins and real investment in their growth, fear turns into engagement.

To reinforce a culture of open learning, encourage teams to share wins and failures using frameworks like *FLOW* and SAR (Situation, Action, Result). One client gave a weekly bonus for the biggest lesson learned from a mistake. It made experimentation fun and safe.

When you apply the Beginner's Playbook, the outcome is clear: Curiosity replaces fear and **communication becomes a strategic accelerator.**

Everyone Plays, From Leadership to Interns

Executives are often the last to notice what's changing; they're busy running what exists. Meanwhile, edge employees are already experimenting.

The people closest to the work see change first. The people closest to power see it last. Leaders must learn to listen down the organization. That's where the first signals, and the first opportunities, appear.

Finding and Empowering Internal Champions

Every organization already has people who are ahead of the curve. They are your Explorers. You do not always need to hire them; most already work for you. Your job is to identify them, elevate them, and make their impact contagious

1. Identify employees talking about or using new tech.
2. Elevate them as **Explorers.** Give them a map (the *FLOW* process) and permission to run tests, share results, and invite others to participate.
3. Equip them with resources such as time, access, and psychological safety. Nothing kills momentum faster than starving enthusiasm.

4. Reward learning, not just winning. Make progress visible and celebrated..

Some companies move Explorers across departments, where they can teach others how to run tests and identify new problems worth solving. One Explorer with space to lead can shift an entire organization. As their excitement spreads, **learning becomes contagious.**

From FOMO to *FLOW*

During times of change, most companies oscillate between two fears: the fear of missing out and the fear of jumping in. They start dozens of projects, overcommit resources, and exhaust people.

FLOW creates calm momentum: **Focus • Learn • Observe • Win and Widen.**

- **Focus:** Start narrow. Pick one problem worth solving.

- **Learn:** Run a short test and share discoveries openly.

- **Observe:** Track and share where value appears and document what you learn.

- **Win and Widen:** Scale what works and retire what doesn't.

FLOW turns innovation chaos into continuous progress. That is a rhythm an entire company can follow.

Transformation isn't about doing more. It's about learning faster.

Shoshin Practice #3: Spot the Open Door

Ask yourself three questions:

1. Where are successful small tests already happening and how can we amplify them?

2. Who is experimenting on their own time and how can we give them a stage?

3. What's one small, low-risk bet we can make this week to learn faster than our competitors?

Opportunity doesn't come from predicting the future. It comes from learning faster than everyone else.

In the next chapter, we'll look at the system that makes this momentum repeatable: *FLOW*, the heartbeat of transformation.

CHAPTER 4

SYSTEM: The *Shoshin* Method (*FLOW*)

How to Turn Momentum into a Repeatable Operating Rhythm

Every Movement Needs Rhythm

Momentum is fragile without rhythm.

Transformation rarely fails because people stop caring. It fails because they don't know what to do next. Most organizations run on adrenaline: a big launch, a big win, a big crash. Then, everything stalls.

The best companies trade adrenaline for rhythm. They don't rely on inspiration; they rely on cadence.

FLOW is that cadence. It turns curiosity into a repeatable operating system. It converts experiments into learning, learning into progress, and progress into a culture that keeps moving forward.

Shoshin gives you the **mindset to start.** *FLOW* gives you the **framework to continue.** Together, they turn transformation from an event into a habit.

Why a Philosophy Needs a Framework

Mindset without structure drifts. Structure without mindset stagnates.

High-performing organizations balance exploration with execution. Shoshin sparks exploration by helping you stay open and ask better questions. *FLOW* provides the discipline to execute, refine, and scale the answers.

The core equation holds true:

Curiosity x Speed to Learning x Implementation = Transformation

Curiosity ignites movement. Speed to learning compounds it. Implementation locks it in place.

Without a framework, enthusiasm fades after the first few wins. With *FLOW*, learning becomes continuous, visible, and measurable.

Introducing the *FLOW* Model

Think of *FLOW* as the gearbox of transformation. **It's simple, predictable, and endlessly repeatable.**

Here is the *FLOW* model in its simplest form.

THE *FLOW* MODEL

Letter	Principle	Description
F	Focus	Choose one clear, high-impact target. Remove every distraction. Multitasking kills momentum.
L	Learn	Share progress openly. Document what works and what does not. Visibility fuels accountability and collective discovery.
O	Observe	Run controlled tests. Track data. Adjust quickly. Replace assumption with evidence.
W	Win and Widen	Scale what works. Retire what does not. Spread learning so each win feeds the beginning of the next *FLOW* cycle.

FLOW is not linear; it loops. Each Win fuels the next Focus. Each cycle compounds learning. Over time, progress becomes predictable.

THE FLOW LOOP

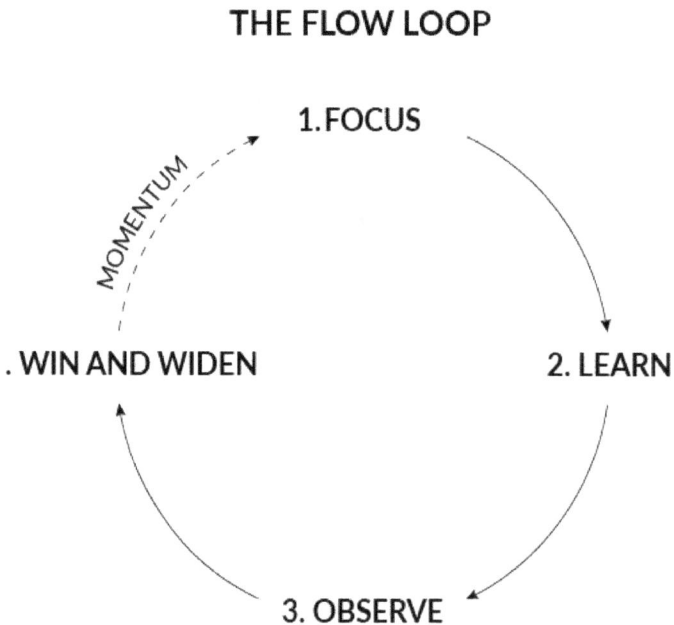

MOMENTUM

1. FOCUS

2. LEARN

3. OBSERVE

. WIN AND WIDEN

Focus: Hammers and Nails Syndrome*

The biggest mistake leaders make is trying to drive too many priorities at once.

Let's say I hand you one hammer and one nail. Your job is simple: Pull the hammer back and drive that nail straight through a piece of wood. You may miss the nail the first or second time. But eventually, you will drive it through.

Now, let's say I give you 10 hammers and 10 nails. Coordination collapses. You can't hit a thing. Eventually, you stop swinging.

This is how most companies operate. They launch multiple initiatives, chase too many priorities, test half-formed pilots, and wonder why nothing sticks.

The truth is simple:

You cannot drive 10 nails at once. And you cannot drive 10 strategic initiatives at the same time and expect progress.

Pick one hammer, one nail, and drive it cleanly. Then move to the next.

That is focus.

That is rhythm.

That is how **transformation becomes real**.

Case Study: From Chaos to Cadence

A global sales organization was drowning in innovation chaos: Twenty pilots. Ten project managers. Zero cohesion. Everyone was "doing AI," but no one was learning from anyone else.

We introduced *FLOW*.

Step 1: Pick one target.

AI-assisted proposal generation. Sales teams were spending hours writing and customizing proposals. The target was clear.

Step 2: Form a small, contained team.

A single sales unit ran the first test.

Step 3: Run a 90-day FLOW cycle.

Each week, the team tracked proposal turnaround time, client feedback, and win rates. They documented what broke, what worked, and what surprised them. At the end of 90 days, they shared the results.

Proposal creation time dropped 35%. Response quality improved. Customer satisfaction climbed.

But the real impact happened next.

Business development realized they could use *FLOW* for follow-up emails and meeting summaries. Marketing saw opportunities for campaign briefs. Teams began comparing notes, swapping templates, and sharing dashboards.

Fear dropped. Confidence rose. Curiosity spread.

Leadership did not force breakthroughs from the top down. They created space for them to rise from the bottom up. Their role was to set tempo, protect culture, and celebrate learning.

Chaos became cadence.

And *FLOW* became **the rhythm the entire organization followed.**

Momentum Math Revisited

In Chapter 1, we learned that small, consistent actions compound over time. In practice, momentum isn't math—it's rhythm. It's what happens when **learning cycles shorten, feedback loops tighten, and every win fuels the next.**

Momentum is built, not found. It comes from a system that keeps moving, even when results are still catching up.

Disciplined teams outperform reactive ones because momentum turns motion into progress and progress into transformation.

FLOW Cadence:

The Rhythm That Sustains Transformation

Every system needs a clock. *FLOW* runs on a simple, repeatable rhythm that keeps learning visible, momentum measurable, and progress continuous.

1. **15-Minute Team Check-Ins**

   Quick, focused, and rhythmic. One question per letter: What are we Focusing on? What did we Learn? What did we Observe? What did we Win?

2. **60-Day Comprehensive Team Reviews**

   Teams review data, capture lessons, remove failed tests, and formalize learning. Short-term insights become institutional knowledge.

3. **90-Day Enterprise-Wide Sharing and Collaboration**

   Departments share results, insights, and templates across the organization. Local experiments become enterprise-wide momentum. This is how transformation scales.

   This cadence is a default rhythm, not a rigid rule. Teams can compress or extend cycles as long as learning remains visible and momentum stays intact.

Leadership's Role in Sustaining *FLOW*

Leadership sets the tempo. When leaders model curiosity, teams match it.

During transformation cycles, the best question an executive can ask is, **"What did we learn this week?"** not, "Did we hit the target?"

This single shift rewires culture. It makes experimentation safe. It rewards progress over perfection. By replacing traditional performance reviews with learning reviews, performance rises naturally.

Leaders don't conduct the orchestra; they keep the beat.

Shoshin Practice #4: Run Your First *FLOW* Sprint

Try this 90 day experiment to install *FLOW* in your organization:

1. **Identify one problem worth solving.**

   Start small with one workflow or customer experience issue.

2. **Form a team inside your department.**

   Have everyone run the same small test and share results. Then, expand to cross-functional teams.

3. **Run the full *FLOW* cycle.**

   Focus: Define goal and success metrics.

   Learn: Hold brief check-ins every other week.

   Observe: Review data during sixty-day team reviews.

   Win and Widen: Present results during the ninety-day enterprise session.

4. **Share learning publicly.**

   Host a 60 minute presentation at the end of the cycle.

5. **Celebrate curiosity.**

   Recognize effort and learning, not just outcomes.

Rhythm as Competitive Advantage

Transformation without rhythm is chaos.

Rhythm without curiosity is repetition.

When Shoshin drives *FLOW*, progress becomes inevitable.

The companies that build rhythm build resilience. Their advantage isn't technology. It's **tempo.**

In the next chapter, we'll look at proof: organizations in every industry applying these principles and what their results reveal.

CHAPTER 5

HIGHLIGHTS:
Proof That the System Works

Ten Stories of Transformation Across Industries

Why Proof Matters

Ideas inspire. Proof persuades.

By now, you understand the mindset of *Shoshin* and the structure of *FLOW*. But leaders don't invest in theory. They invest in evidence. They want to know one thing: **Does it work?**

This chapter answers that question through 10 concise case studies drawn from real organizations that applied the same rhythm:

Small Tests → Visible Learning → Shared Wins

Transformation is not industry-specific. It's human-specific. Any organization willing to start small, stay curious, and keep moving can replicate these results.

THE SAR CASE STUDY TEMPLATE

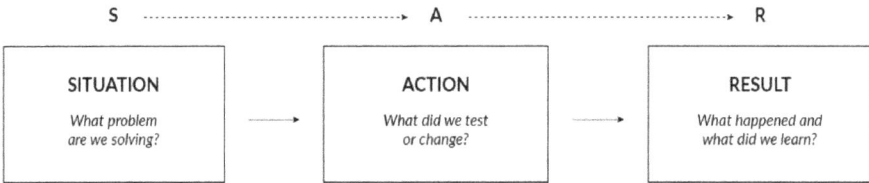

SITUATION	ACTION	RESULT
What problem are we solving?	What did we test or change?	What happened and what did we learn?

1. Health Care: From Overload to Insight

Situation: A hospital network was drowning in documentation. Physicians spent more time typing than treating.

Action: A 90-day *FLOW* sprint automated intake summaries and anomaly flags through a secure AI model.

Result: Documentation time dropped 35%, accuracy rose 20%, and clinician satisfaction surged.

2. Finance: Closing the Books in Half the Time

Situation: Month-end close required 10 days.

Action: The CFO approved a micro-pilot using an AI reconciliation tool on one account. Once it worked, the team scaled it firm-wide in six weeks.

Result: Close time fell to five days. Audit accuracy increased by 10%. Finance became a transformation leader.

3. Manufacturing: Predict, Don't React

Situation: Reactive maintenance eroded margins.

Action: Installed sensors on one production line and ran a *FLOW* cycle.

Result: Unplanned downtime decreased by 22%. Inventory costs dropped by 9%. Firefighting became forecasting.

4. Media: Scaling Creativity Without Losing Voice

Situation: Content demand outpaced team capacity.

Action: Writers used AI for first drafts. Editors refined tone. Results posted weekly in a shared channel.

Result: Output doubled. Engagement rose 14%. Posts per asset dropped 30%. AI handled the volume; humans kept voice.

5. Education: Teaching Transformation

Situation: Corporate learning couldn't keep up with technological change.

Action: Launched an internal *Shoshin* Academy teaching AI fundamentals and *FLOW*.

Result: 80% of graduates led measurable innovations within 90 days. Learning became a leadership path.

6. Retail: From Forecast to *FLOW*

Situation: Marketing and supply-chain teams worked in silos. Promotions outpaced inventory.

Action: Created a cross-functional "one nail" team running weekly *FLOW* reviews.

Result: Forecast accuracy rose 18%. Markdowns decreased 12%. Trust between departments was restored.

7. Logistics: Finding Efficiency in the Gaps

Situation: Delivery routes were inconsistent and fuel costs rising.

Action: Piloted AI-based route optimization in one region. Drivers shared observations daily.

Result: On-time deliveries increased 25%. Fuel cost decreased 8%. Employee turnover declined 10%.

8. Human Resources: Building a Culture of Curiosity

Situation: Employees feared AI would replace them.

Action: Formed cross-department "Explorer Teams" to test tools and share lessons. Introduced a "Best Experiment of the Week" award.

Result: Tool adoption rose 65%. Engagement index jumped by 20 points. Fear became fascination.

9. Legal & Compliance: From Bottleneck to Bridge

Situation: Legal slowed every initiative due to risk concerns.

Action: General Counsel joined early and co-created AI governance templates.

Result: Review times dropped 30% with no increase in risk. Legal became an enabler.

10. Health Network: Scaling Curiosity

Situation: A national provider wanted to replicate results across its hospital network.

Action: Standardized 90-day *FLOW* cycles across regions and shared wins through a "Learning Map" dashboard.

Result: $40 million annual savings. Employee engagement at record highs. Curiosity became system wide.

Debrief: The Patterns Behind the Proof

The obstacles were identical: fear, ego, silos, inertia. Each organization overcame them by learning faster than fear could spread.

Transformation isn't an accident. It's a system.

PATTERNS BEHIND THE PROOF

Trait	Description
Speed	Small cycles solved big problems.
Inclusion	Everyone participated, from interns to executives.
Visibility	Shared wins multiplied learning.
Mindset	Curiosity consistently outperformed expertise.

The obstacles were identical: fear, ego, silos, inertia. Each organization overcame them by learning faster than fear could spread.

Transformation isn't an accident. It's a system.

Shoshin Practice #5: Extract the Lesson

Use the SAR model to turn experience into reusable intelligence.

1. **Situation:** What problem were we solving?
2. **Action:** What did we test or change?
3. **Result:** What happened and what did we learn?

Record, repeat, refine.

From Proof to Practice

Every success in this chapter started with a question, not an answer. Each one turned curiosity into rhythm and rhythm into measurable progress.

Opportunity is what beginners see before experts can explain why it won't work. It appears when courage meets curiosity.

The evidence is clear: ***Shoshin* and FLOW aren't theories. They're a system that delivers.**

In the next chapter, we'll move from proof to implementation and explore how to build an AI-ready organization and make transformation everyone's job.

CHAPTER 6

IMPLEMENTATION:
Building an AI-Ready Organization

How to Turn Insight into Execution and Make Transformation Everyone's Job

From Proof to Process

Transformation only lasts when it becomes structure.

You can have the right mindset, the right model, and even the right momentum. But without implementation, progress fades. Ideas remain ideas. Wins remain isolated. And transformation remains a project instead of becoming culture.

This chapter turns the *Shoshin*-OS "on." It shifts *Shoshin* and *FLOW* from philosophy to operating model. Becoming an AI-ready organization isn't about adding new tools. It's about rewiring how people think, work, and learn together.

The goal is simple:

Make transformation everyone's responsibility, every day.

The *Shoshin* Operating System (*Shoshin*-OS)

IMPLEMENTATION
HOW TO BUILD AN AI-READY ORGANIZATION

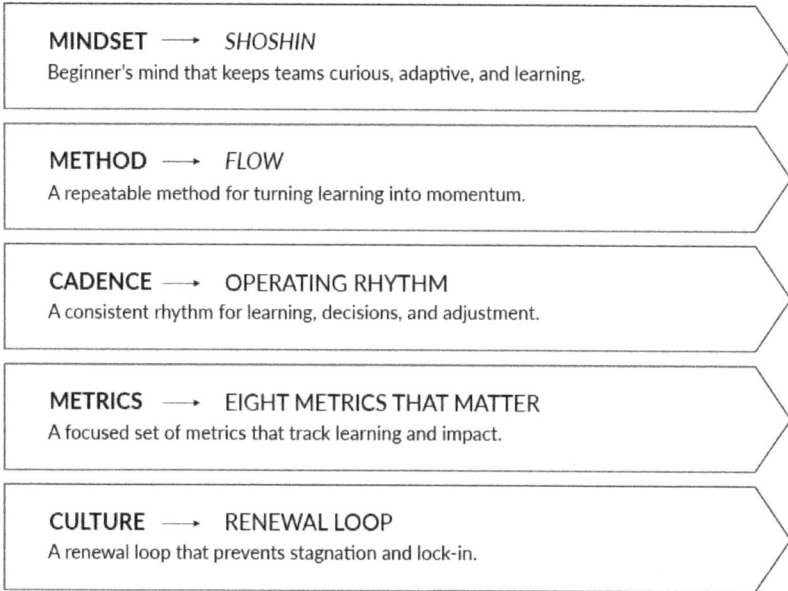

MINDSET ⟶ *SHOSHIN*
Beginner's mind that keeps teams curious, adaptive, and learning.

METHOD ⟶ *FLOW*
A repeatable method for turning learning into momentum.

CADENCE ⟶ OPERATING RHYTHM
A consistent rhythm for learning, decisions, and adjustment.

METRICS ⟶ EIGHT METRICS THAT MATTER
A focused set of metrics that track learning and impact.

CULTURE ⟶ RENEWAL LOOP
A renewal loop that prevents stagnation and lock-in.

Turning Proof into Process

High-performing organizations do not just experiment. They **operationalize** transformation.

The *Shoshin* Operating System aligns the five drivers of change into one repeatable framework:

1. **Mindset**

2. **Method**

3. Cadence

4. Metrics

5. Culture

When these five elements work together, the Shoshin-OS stops transformation from being episodic and turns it into a way of operating.

1. Mindset: *Shoshin*

Shoshin is the foundation. It is the beginner's mind that fuels curiosity, openness, and continuous exploration. Without Shoshin, transformation becomes compliance. With Shoshin, it becomes a possibility.

Shoshin is the belief that something better is always possible, and the discipline to keep asking questions until you find it..

2. Method: *FLOW*

FLOW is the four-step rhythm that turns learning into progress.

FLOW FOUR-STEP RHYTHM

Letter	Principle	Description
F	Focus	Choose one clear, high-impact target. Remove every distraction. Multitasking kills momentum.
L	Learn	Share progress openly. Document what works and what does not. Visibility fuels accountability and collective discovery.
O	Observe	Run controlled tests. Track data. Adjust quickly. Replace assumption with evidence.
W	Win and Widen	Scale what works. Retire what does not. Spread learning so each win feeds the beginning of the next *FLOW* cycle.

Shoshin creates curiosity. *FLOW* creates structure. When you combine them, you get repeatable momentum.

3. Cadence: The Rhythm That Sustains Transformation

Mindset and method need rhythm.

Cadence ensures learning stays consistent, visible, and predictable across the enterprise.

15-Minute Check-Ins (Every Other Week)

Teams answer four questions:

1. What are we Focusing on?

2. What did we Learn?

3. What did we Observe?

4. What did we Win and how can we Widen?

60-Day Team Reviews

Teams evaluate progress, retire failed tests, and capture lessons.

Short-term insights become institutional knowledge.

90-Day Enterprise Collaboration Sessions

Teams share results, templates, and insights across the company.

Local wins become enterprise-wide momentum.

Cadence builds the heartbeat of transformation.

4. Metrics: The Eight Metrics That Matter

What gets measured gets multiplied.

These eight metrics help leaders track transformation health across each 90 day cycle:

1. **Education Speed**
   How quickly employees build AI fluency

2. **Adoption Rate**
   Percent of employees using AI tools or *FLOW* cycles

3. **Speed to Learning**
   Time from test to insight

4. **Experiment Velocity**
   Number of tests completed each quarter

5. **Cross-Functional Participation**

   Breadth of departments involved

6. **Efficiency Lift**

   Time or cost savings produced during each cycle

7. **Engagement Index**

   Employee sentiment toward innovation

8. **Implementation Rate**

   Percent of successful tests scaled enterprise-wide

Publish these metrics on a Transformation Dashboard.

Visibility creates pride. Pride sustains momentum.

5. Culture: The Renewal Loop

Culture is the ultimate multiplier.

The Renewal Loop keeps the organization from stagnating by creating a simple sequence:

Observe → Adapt → Reinforce → Restart

Culture becomes a living system where learning never stops and reinvention becomes routine.

When mindset, method, cadence, metrics, and culture align, transformation becomes the operating system of the organization—

not an initiative.

The 90-Day Shoshin Sprint

The Shoshin Sprint compresses learning, alignment, and execution into 90 days. Each sprint follows the *FLOW* rhythm:

Focus • Learn • Observe • Win and Widen.

THE SHOSHIN SPRINT

Phase	Theme	Key Action
Weeks 1–3	Focus	Select one meaningful business challenge. Define success metrics and an accountable owner.
Weeks 4–8	Learn and Observe	Run two or three small tests in parallel. Share insights during 15-minute check-ins.
Weeks 9–12	Win and Widen	Measure impact, delete what does not work, and scale what does. Prepare enterprise-wide sharing.

Restart every 90 days.

Repetition compounds momentum.

Roles, Rhythms, and Rituals

Clarity Prevents Chaos

Clear roles prevent drift and confusion.

Leadership

Set vision. Model curiosity. Protect time for experimentation.

Managers

Facilitate learning cycles. Remove barriers. Track insights.

Teams

Run tests. Collect data. Share learning openly.

Explorers

Drive experimentation, connect departments, mentor peers.

When everyone knows their role in the system, progress accelerates.

Turning Play into Process

Play sparks innovation. Process sustains it.

To convert exploration into repeatable progress:

- Build an internal AI Library of validated use cases

- Create simple templates for experiments

- Launch a Community of Practice where teams share discoveries

- Celebrate learning, not just outcomes

When play becomes a process, transformation becomes a habit.

Investing in the Tools of Transformation

There is one more element leaders often overlook: **access.**

Learning at the speed of AI requires access to the tools that power experimentation. ChatGPT, Claude, Gemini, and Copilot cost as little as others spend on lunch.

When leaders restrict access, employees experiment privately, results stay hidden, and learning slows.

In the future, transformation-ready organizations will treat AI tools as essential infrastructure, as important as laptops or internet access.

If you expect people to drive innovation, you must equip them to do it.

Otherwise, your most curious employees will leave for companies that do.

Measuring Momentum: From Pilot to Practice

A professional-services firm began with one AI pilot. Early success led leadership to form a permanent *Shoshin* Team running continuous 90 day cycles across divisions.

Each quarter they trained new Explorers, launched tests, and published outcomes publicly.

After one year:

- 80% of departments ran at least one *FLOW* cycle

- Average project cycle time decreased 28%

- Engagement scores reached all-time highs

Transformation became muscle memory. It became a way of working, not a project.

Shoshin Practice #6: Your 90-Day Implementation Plan

1. **Pick one project worth testing.**

2. **Run the test inside your department with at least one Explorer.**

3. **Define goals, roles, metrics, and milestones.**

4. **Run a full 90 day *FLOW* cycle:**

    Bi-weekly: 15-minute check-ins

    60 days: Reflect, refine, retire

    90 days: Enterprise sharing and collaboration

5. **Document learning and share it openly.**

    Implementation isn't about size. It's about the system.

The Momentum Multiplier

Once transformation becomes everyone's job, it stops being anyone's project. It becomes culture.

Momentum multiplies when people, systems, and incentives align. It is no longer powered by urgency. It is powered by rhythm.

Momentum is not built by machines. It is built by people who stay curious.

In the next chapter, we explore how to institutionalize that rhythm so transformation becomes the permanent state of your organization.

CHAPTER 7

NEVER STOP:
Institutionalizing Transformation

How to Build a Culture That Keeps Beginning

The Beginner's Habit

Transformation isn't an event. It is a rhythm.

Most organizations treat change like a campaign: plan, launch, celebrate, move on. But in a world where markets, technology, and customer expectations shift constantly, there is no finish line. The advantage belongs to companies that make beginning again a permanent habit.

Shoshin **introduced the mindset.** *FLOW* **created the operating rhythm. This chapter shows how to turn both into culture.**

The goal is not endless reinvention. The goal is the discipline of renewal. Great companies do not ask, **"How do we change?"** They ask, "How do we keep changing?"

Here's how that rhythm looks when you institutionalize it.

The Renewal Loop

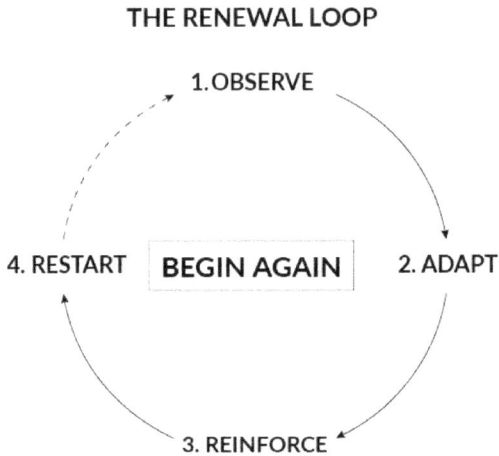

THE RENEWAL LOOP

```
              1. OBSERVE



4. RESTART   | BEGIN AGAIN |   2. ADAPT



              3. REINFORCE
```

The system you've built in Chapter 6 is what makes this next step possible. Now, we move from system to culture, from rhythm to renewal.

The **Renewal Loop** keeps transformation alive. It's the cycle that prevents curiosity from fading into comfort. Visualize it as a circle, not a line, with each turn feeding the next.

THE RENEWAL STAGES

Stage	What Happens
Observe	Scan the environment for new signals such as emerging technology, customer shifts, and employee insights.
Adapt	Translate signals into small, fast tests.
Reinforce	Share wins, embed what works, and reward the people who contributed.
Restart	Ask new questions and begin a new cycle.

Organizations don't get stuck because they run out of ideas. They get stuck because they stop asking new questions.

Netflix reinvented itself four times using this pattern: DVDs to streaming, licensing to producing, human curation to algorithms, and now gaming. Each loop of observation, adaptation, and reinforcement created the next breakthrough.

That's *Shoshin* institutionalized.

Building Organizational Shoshin

A *Shoshin* culture rests on four daily habits: openness, experimentation, reflection, and renewal.

1. **Hire for curiosity, not just credentials.**
    The best employees are the ones who keep asking why.

2. **Promote for adaptability.**
    Reward people who pivot quickly when the data changes.

3. **Measure learning velocity.**
    Track how fast teams move from Question → Insight → Action.

4. **Encourage learning sabbaticals.**
    Give people structured time to explore, then return to teach.

One global tech firm now gives every manager a learning KPI. Each quarter, they must present one new insight or skill that benefits the business. It's simple and effective.

Culture doesn't shift because you announce new values. Culture shifts because people see leaders living them.

Quarterly "Start Over" Sessions

Every 90 days, pause and ask:

"If we were starting today, what would we do differently?"

These sessions institutionalize humility. Cross-functional teams review priorities through the *FLOW* lens and decide what to keep, fix, or retire.

This is the corporate equivalent of meditation—a deliberate reset before momentum turns into complacency.

Because the greatest threat to transformation isn't failure. It's success.

The Superhero Trap

Every organization has a superpower.

It's the thing that made you successful in the first place. The capability, model, or advantage that separated you from everyone else and carried you through your last era of growth.

And that is precisely the problem.

In stable environments, superpowers **compound.** In changing environments, they **calcify.**

Blockbuster mastered physical distribution. Kodak mastered film chemistry. Taxi companies mastered scarcity through medallions. Each advantage worked brilliantly—until the environment shifted. What once created dominance slowly became drag.

This is the Superhero Trap: the moment when yesterday's strength becomes today's blind spot.

The trap is subtle because it feels like discipline. Leaders defend the system that worked. Teams optimize what they know. Metrics reward efficiency instead of exploration. Over time, identity hardens around success.

The organization stops asking beginner questions. Not because it can't, but because it doesn't feel necessary. This is how great companies lose curiosity.

They don't fail at **execution.**

They fail at **renewal.**

In the age of AI, the Superhero Trap accelerates. Capabilities that took years to build can be replicated or replaced in months.

The danger isn't **pride.**

The danger is **certainty.**

That's why renewal cannot rely on leadership intuition or occasional strategy offsites. It must be institutionalized.

Quarterly "Start Over" sessions exist to break the Superhero Trap on purpose. They force the organization to ask:

- *If we were starting today, would we still build it this way?*
- *Which strengths are helping us and which are quietly holding us back?*
- *What are we defending out of habit instead of evidence?*

This isn't about abandoning strengths. It's about preventing them from becoming constraints.

The most enduring organizations treat humility as a system. They design moments where identity is loosened, assumptions are challenged, and beginner's mind is reactivated—before the market forces them to do it under pressure.

Shoshin isn't the absence of expertise.

 It's the refusal to be **trapped by it.**

Renewal is the discipline of letting go early—while belief is still intact, momentum is still strong, and choice is still available.

Continuous Reinvention: Staying Ahead of the Curve

Transformation doesn't stop at implementation. It loops. Companies that stay relevant use renewal as strategy.

- **Revisit the Transformation Bridge:** Keep Initiators and Translators visible. Bring Builders along faster.
- **Track early signals:** Watch customers, employees, and startups for emerging behavior.

- **Build external partnerships:** Collaborate with universities and labs for fresh perspectives.

- **Leverage AI dashboards:** Monitor innovation trends and internal idea flow.

The goal isn't to avoid disruption; it's to **make disruption your advantage.**

The Human Advantage

Technology gives leverage. People give meaning.

AI can summarize, simulate, and predict, but it can't care. The enduring differentiator is human curiosity, creativity, and empathy. **When people feel safe to explore and question, innovation multiplies.**

A logistics CEO told me, "Automation didn't replace our people; it freed them to solve problems we didn't know we had."

Technology accelerates progress. People sustain it.

Shoshin Practice #7: The Renewal Loop

Use this quarterly to keep your organization fresh:

1. **Observe:** Identify new signals in tech, markets, and behavior.
2. **Adapt:** Turn one signal into an experiment.
3. **Reinforce:** Share learning and celebrate contributors.

4. **Restart:** Ask "What's next?" and begin again.

Renewal is simply the discipline of asking again what beginners ask naturally:

Why? What if? What next?

The Infinite Game

There is no final stage of transformation, only the next beginning.

In a world where change never stops, the greatest leaders are the ones who never stop beginning.

Shoshin isn't the end of transformation; it's the practice that keeps it alive. The organizations that thrive in the age of AI will need both technology and curiosity.

Transformation is never finished. **The moment you stop learning, you stop leading.**

The work you have done through the first seven chapters built the mindset, the method, and the rhythm of transformation. But systems only create change when they are practiced. The next step is to put *Shoshin* and *FLOW* to work inside your organization.

The Playbook that follows gives you the tools, templates, and exercises to turn the system into daily action.

PART II:

The *Shoshin* Playbook

THE *SHOSHIN* PLAYBOOK

Tools, Templates, and Systems for Continuous Transformation

Support for Your Journey to AI-Powered Business Transformation

Part I gave you the *Shoshin* Operating System: the mindset, method, and rhythm that drive transformation.

Part II gives you the tools to put it into practice.

This Playbook contains the exercises, templates, and workflows that leadership teams use to build AI-ready organizations. Each Practice includes a QR code linking to digital worksheets so your teams can collaborate, share results, and track momentum over time.

Use the Playbook to run your next 90-day cycle, install FLOW inside your teams, and build the habits that make transformation continuous.

Shoshin is the mindset.

FLOW is the system.

This Playbook is where you begin.

THE 90-DAY *SHOSHIN* CYCLE

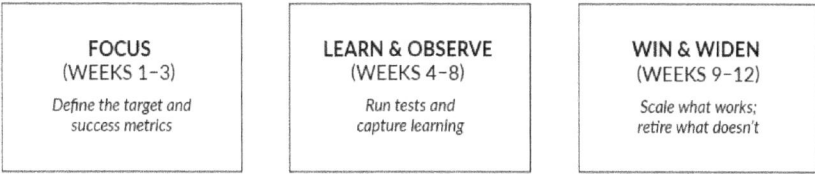

FOCUS (WEEKS 1–3)	LEARN & OBSERVE (WEEKS 4–8)	WIN & WIDEN (WEEKS 9–12)
Define the target and success metrics	*Run tests and capture learning*	*Scale what works; retire what doesn't*

Practice #1: Start Small, Start Now (The Daily Reset)

Chapter Alignment: Chapter 1: START

<u>Purpose</u>

To reset your mindset each day, surface assumptions, and build momentum through small actions.

<u>Instructions</u>

Spend 10-15 minutes each morning answering the questions below. Capture insights weekly.

Step 1: Examine Core Beliefs

What beliefs about our business have served us well? Do they still serve us today?

- List 3-5 long-standing beliefs.

- Mark each as: Still True / Needs Updating / No Longer True.

Step 2: Apply the "Day One" Lens

If I were starting this company or department from scratch today, what would I build differently?

- Identify 2-3 processes, habits, or assumptions you would not bring forward.

- Identify 1-2 new opportunities you'd act on immediately.

Step 3: Choose One Micro-Action

What is one small action I can take right now to build momentum?

- Choose something achievable in under 20 minutes.

- Examples: Refine a prompt, test one workflow, clean one dataset, rewrite one SOP, share one insight.

Step 4: Hold a Weekly Reflection

At the end of each week, ask yourself:

- What patterns emerged?

- What did I question or let go of?

- What changed because of one small daily action?

Practice #2: See the Pattern

Chapter Alignment: Chapter 2: HISTORY

<u>Purpose</u>

To help leaders identify early signals of transformation and avoid being trapped by past success.

<u>Instructions</u>

Spend 10-15 minutes each week answering the questions below. Capture insights monthly.

Step 1: Identify What You're Holding Onto

What are we holding onto simply because it has always worked?

- List 5 legacy processes, products, or assumptions.

- Mark each as: Keep / Evolve / Sunset.

Step 2: Identify Internal Explorers

Who already sees the next wave and is experimenting?

- List 3-5 individuals or teams already testing new tools or approaches.

- What are they noticing that others are not?

- How can you amplify their learning?

Step 3: Define Your Proof Points

What proof points can we create this quarter to help others cross the Transformation Bridge?

- Choose 1-2 tests that demonstrate value quickly.

- Define success using measurable outcomes (i.e., speed, quality, cost, satisfaction, accuracy).

Step 4: Review the Pattern

Every month, ask:

- What signals appeared?

- What assumptions changed?

- What new opportunities are visible now?

Practice #3: Spot the Open Door

Chapter Alignment: Chapter 3: OPPORTUNITY

Purpose

To help teams identify small early wins that create momentum and reduce fear.

Instructions

Step 1: Find Existing Small Tests

Where are successful small tests already happening?

- Identify 3-5 informal experiments already occurring.

- What value have they created?

Step 2: Identify Independent Experimenters

Who is experimenting on their own time?

- List early adopters.

- What tools are they using?

- What did they learn that others should know?

Step 3: Create One Low-Risk Bet

What is one small, low-risk bet we can make this week to learn faster than competitors?

- Define the test.

- Identify owners.

- Define success metrics (keep it simple: time saved, errors reduced, clarity gained).

Step 4: Share It

Document and share the learning internally within one week.

Practice #4: Run Your First *FLOW* Sprint

Chapter Alignment: Chapter 4: SYSTEM

<u>Purpose</u>

To install the *FLOW* rhythm in a single team over a 90-day cycle.

<u>Instructions</u>

Step 1: Identify One Problem Worth Solving

- Choose one workflow that impacts speed, cost, or quality.

- Define the expected business outcome.

Step 2: Form Your Team

- Include at least one Explorer.

- Assign roles: Lead, Analyst, Documenter, Communicator

Step 3: Run the 90-Day FLOW Cycle

- Focus: Define the goal, success metrics, and guardrails.

- Learn: Run tests. Hold 15-minute check-ins every other week.

- Observe: Review data during the 60-day review.

- Win and Widen: Present results in the 90-day sharing session.

Step 4: Share Openly

- Conduct a 60-minute debrief.

- Capture insights.

- Publish the template and results for others to reuse.

Practice #5: Extract the Lesson

Chapter Alignment: Chapter 5: HIGHLIGHTS

Purpose

To turn experience into reusable intelligence using the SAR model.

<u>Instructions</u>

Step 1: Define the Situation

- What problem were we solving?

- What was the context?

- What was at stake?

Step 2: Describe the Action

- What did we test or change?

- Who did what?

- What was the scope?

Step 3: Capture the Result

- What happened?

- What surprised us?

- What would we do differently next time?

Step 4: Publish the SAR

- Share with your team.

- Add to your internal AI Library.

- Use it to guide future tests.

Practice #6: Your 90-Day Implementation Plan

Chapter Alignment: Chapter 6: IMPLEMENTATION

<u>Purpose</u>

To scale *FLOW* beyond a single team and turn it into organizational rhythm.

<u>Instructions</u>

Step 1: Pick One Project to Test

- Choose a process that creates friction or delay.

- Define the business impact.

Step 2: Build Your Cross-Functional Team

- Include at least one Explorer.

- Define roles.

- Define communication cadence.

Step 3: Run a Full 90-Day Cycle

- **Biweekly:** Check-ins

- **60-day:** Review + reflect + refine

- **90-day:** Present results + templates

Step 4: Document and Share

- Upload results to your internal AI Library.

- Publish dashboards.

- Share what worked and what didn't.

Step 5: Repeat

Start next quarter's iteration immediately.

Practice #7: The Renewal Loop

Chapter Alignment: Chapter 7: NEVER STOP

Purpose

To institutionalize learning, renewal, and reinvention.

Instructions

Step 1: Observe

- Identify new signals in markets, tech, customer behavior, employee ideas.

Step 2: Adapt

- Turn one signal into a simple experiment.

- Define owner, scope, and desired outcome.

Step 3: Reinforce

- Share the learning.

- Embed what worked into standard practice.

- Recognize contributors.

Step 4: Restart

- Ask "What's next?"

- Start the next Renewal Loop immediately.

Download The Shoshin Playbook

Scan the QR code below to download the full Shoshin Playbook and other digital worksheets your team can use to collaborate and track progress.

Gratitude & Acknowledgments

To every leader who has shared their Shoshin stories, tested FLOW inside their companies, and helped refine these frameworks in real time—thank you.

To my daughters, Lily and Lexi - you inspire everything I do.

To the people who stuck with me and never gave up during every transformation: My mom and dad - Lisa and Don Sinclair, Ashley and Shane Maidy, Jason Reid, Greg Reid, Erika Persily, and Dan Clark.

And to every reader who dares to ask, "What's next?" - this book was written for you.

References

- Scott Duffy, *Breakthrough: How to Harness the Aha! Moments That Spark Success* (Entrepreneur Press, 2018). * Used with permission.

- Geoffrey A. Moore, *Crossing the Chasm* (Harper Business, 1991).

- Shunryu Suzuki, *Zen Mind, Beginner's Mind* (Weatherhill, 1970).

- Phil Jackson, *Eleven Rings* (Penguin Press, 2013).

- Aditya Challapally, Chris Pease, Ramesh Raskar, and Pradyumna Chari. The GenAI Divide: State of AI in Business 2025 (MIT, 2025). [Report].

- Gartner. Hype Cycle for Emerging Technologies (Gartner, 2025). [Report]

- Greg Reid, The Millionaire Mentor, (Sherpa Press, 2021).

Additional Resources

Transformation does not happen in isolation. The leaders who create the most momentum are the ones who learn in public, share what works, and stay connected to others on the same path.

ScottDuffy.com is your gateway to the broader *Shoshin* community: a place to deepen your practice, access new tools, and stay connected to the evolution of AI-powered business transformation.

Inside the community, you can:

- Learn about Scott's latest AI ventures, keynote speaking, and events
- Access updated tools, templates, and resources to support your next 90 day cycle
- Join programs designed to build AI fluency and develop an AI-ready workforce
- Connect with other leaders applying the Shoshin Operating System inside their organizations

Cultivate Your Beginner's Mind

Scan the QR code below to download the full Shoshin Playbook for your team, join the Shoshin community, and explore other tools and resources to help you sharpen your beginner's mind.

About the Author

Scott Duffy is an entrepreneur and AI business strategist who systematized transformation. Over three decades, he has helped launch and scale companies at the forefront of every major innovation wave—from the Information Age, through the Internet Era, and today's AI Revolution.

Scott is the founder of multiple AI ventures including AI Mavericks, an AI workforce training company and AI Tool Setup, a managed services provider helping AI tool companies scale to the enterprise. Before launching into the AI space, he built a company that was later acquired by Richard Branson's Virgin Group, and held leadership roles at FOXSports.com, NBC Internet, and CBS Sportsline.

Scott began his career working for Tony Robbins, served as a Special Project Editor at Inc.com, co-hosted a popular podcast for Microsoft, and has been recognized by Entrepreneur.com as a "Top 10 Speaker." His insights have been featured by major media outlets including CNBC and he has spoken at the New York Stock Exchange. He is the author of four influential books including *Shoshin: The System for AI-Powered Business Transformation, and The Ultimate Prompting Guide: Your Step-By-Step Guide To Thinking Smarter, Moving Faster, and Achieving More with AI.* Scott is a member of the National Speakers Association.

***

For more information on Scott Duffy or to hire him to speak at your next event, scan the QR code below or visit scottduffy.com.

(Parole ritrovate)

Titolo originario: *La nuova gestione unitaria dell'attività ispettiva: L'Ispettorato Nazionale del Lavoro* / Cristina Giuffrida

In copertina: *Gianni Giuffrida*

Questo libro è stato edito da Zerobook:
 www.zerobook.it.
Prima edizione: 21 febbraio 2018
print: ISBN 978-88-6711-134-3

Controllo qualità ZeroBook: se trovi un errore, segnalacelo!
zerobook@girodivite.it